Outlook for Fish to 2020

Meeting Global Demand

Christopher L. Delgado
Nikolas Wada
Mark W. Rosegrant
Siet Meijer
Mahfuzuddin Ahmed

A 2020 Vision for Food, Agriculture, and the Environment Initiative
International Food Policy Research Institute
Washington, D.C., U.S.A.

WorldFish Center
Penang, Malaysia

October 2003

ISBN 0-89629-647-4

Contents

Preface

The seemingly inexhaustible oceans have proved to be finite after all. Landings of wild fish have leveled off since the mid-1980s, and many stocks of fish are fished so heavily that their future is threatened. And yet the world's appetite for fish has continued to increase, particularly as urban populations and incomes grow in developing countries. Aquaculture—fish farming—has arrived to meet this increased demand. Production of fish from aquaculture has exploded in the past 20 years and continues to expand around the world. But will aquaculture be sufficient to provide affordable fish to the world over the next 20 years? And what environmental and poverty problems will aquaculture face as it expands? Using a global model of supply and demand for food and feed commodities, this report projects the likely changes in the fisheries sector over the next two decades given present trends. As prices for most food commodities fall, fish prices are expected to rise, reflecting demand for fish that outpaces the ability of the world to supply it. Alternative scenarios using different assumptions are also investigated.

The model shows that developing countries will consume and produce a much greater share of the world's fish in the future, and trade in fish commodities will also increase. As aquaculture expands, especially in developing countries, environmental concerns such as effluent pollution, escaped farmed fish, land conversion, and pressure on stocks from fishmeal demand will only increase with time unless technologies and policies promote sustainable intensification. And small, poor producers are at risk of being excluded from rapidly growing export markets unless ways can be found to facilitate affordable certification of food safety and environmentally sound production.

A more complete and detailed version of this report is available in the book *Fish to 2020: Supply and Demand in Changing Global Markets*, by Christopher L. Delgado, Nikolas Wada, Mark W. Rosegrant, Siet Meijer, and Mahfuzuddin Ahmed, also available from IFPRI and the WorldFish Center.

Introduction

The world's fish sector may become a victim of its own success. In the past 30 years the global appetite for fish has doubled. From 45 million metric tons[1] in 1973, total fish consumption jumped to more than 91 million tons in 1997. This enormous growth signals changes in who is consuming fish and where. Consumption of fish in the developed countries stagnated between 1985 and 1997, mainly because populations remained stable and people there were already eating large quantities of fish. But at the same time, rapid population growth in the developing world, along with increases in the average amount of fish consumed per person in those countries, led to soaring increases in global fish consumption.

Now fish production, which has burgeoned to meet rising demand, is subject to growing crises and controversy. The Food and Agriculture Organization of the United Nations (FAO) has repeatedly sounded the alarm over threatened stocks of wild fish and has classified most wild fisheries as either fully exploited or over-exploited.[2] Civil society in developed countries has also become more involved in fisheries issues. A recent report by a panel of scientists, fishers, environmentalists, and policymakers pointed to overfished and depleted stocks in U.S. waters, along with severe habitat degradation.[3]

Aquaculture—the farming in captivity of fish previously caught only in the wild—seems to promise vast new food production resources that can relieve pressure on overburdened wild fisheries and provide a source of fish as food for the poor. Yet aquaculture raises its own problems. Aquaculture development has resulted in massive changes in land use, polluted neighboring waters with effluent, and spread disease among fish farms. Many studies have warned of the potential risks that escaped farmed fish pose for wild populations. And a 2000 study in *Nature* argued that although aquaculture can significantly boost world fish supplies, it also presents sizable environmental tradeoffs by raising demand for wild-caught fish as an ingredient in feed for farmed fish.[4]

So what is the role of aquaculture and wild fisheries in a globalizing food economy, and how will trends in fisheries affect the poor and the environment during the next two decades? This report, based on the book *Fish to 2020: Supply and Demand in Changing Global Markets*, examines changes in the fish sector; the forces driving these changes; and the implications of the changes for fish consumption, production, prices, trade, the environment, and the world's poor. It offers a heretofore missing element in the growing dialog among biologists, ecologists, and policymakers on world fisheries: a broad economic analysis of the rapid changes in fisheries over the past three decades. It looks at fish as a series of market commodities that compete with each other and with meat in food markets and compete with vegetable crops in feed markets. By doing so, it draws explicit attention to the roles of structural changes in demand, policies, and technology in affecting fish prices—and how prices in turn affect both consumer and producer behavior. The study concludes with specific suggestions on entry points for improving the poverty-reduction and environmental impacts of fisheries development.

Recent Trends in Fish Supply and Demand

Since the early 1970s production, consumption, and long-distance trade of fish have risen dramatically, almost entirely because of changes taking place in developing countries. The primary driver of most of these changes has been the increased consumption of fish in developing countries. As population in these countries has grown and consumers have become richer, the resulting increased demand for fish has altered markets for seafood around the world.

Burgeoning Demand for Fish in Developing Countries: The Main Driver of Change

Evidence suggests that the large increases in supply of fish in recent decades have resulted from large increases in demand for fish as food in developing countries. Global consumption of fish has doubled since 1973, and the developing world has been responsible for 90 percent of this growth. Whereas the growth of fish as food in the richer countries has tapered off, in the poorer countries it has grown rapidly. China dominates aggregate consumption of fish products. It accounted for about 36 percent of global consumption in 1997, compared with only 11 percent in 1973 (Figure 1). India and Southeast Asia together accounted for another 17 percent in 1997, with total consumption doubling since 1973. Meanwhile, per capita fish consumption in Sub-Saharan Africa has hardly increased over the past 30 years and in fact has declined since the mid-1980s. Total consumption levels have declined in the developed countries since the mid-1980s, mainly as a result of dramatically

Figure 1 Changing share of developing countries in fish consumption, 1973 and 1997

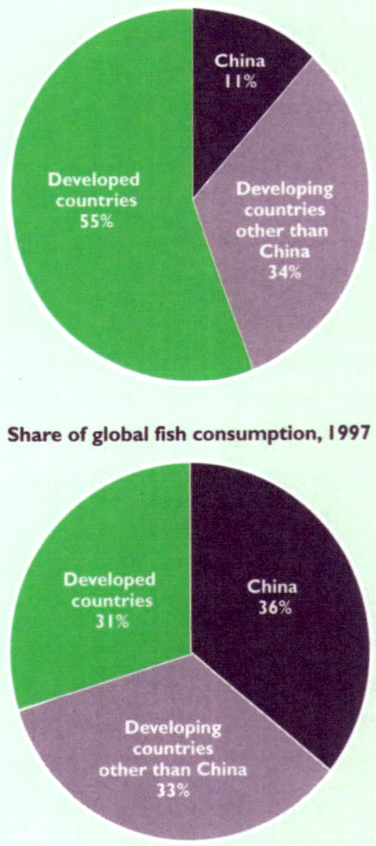

SOURCE: Calculated by authors from FAO statistical databases, www.apps.fao.org/subscriber (accessed January 2002).

NOTE: Data are three-year averages centered on 1973 and 1997, respectively.

Table 1 Total per capita consumption of fish as food, 1973–97

REGION/COUNTRY	TOTAL CONSUMPTION (KG/CAPITA/YEAR) 1973	1985	1997	ANNUAL GROWTH RATE 1985–97 (%)
China	5.5	8.1	26.5	10.4
Southeast Asia	17.6	19.8	23.0	1.3
India	3.1	3.6	4.7	2.3
Other South Asia	6.2	5.4	6.0	0.9
Latin America	7.0	9.0	7.8	–1.2
West Asia and North Africa	3.4	6.2	6.2	0.0
Sub-Saharan Africa	9.0	9.2	6.7	–2.6
United States	13.5	18.5	19.7	0.5
Japan	70.2	61.5	62.6	0.2
European Union 15	18.2	20.3	23.6	1.3
Eastern Europe and former Soviet Union	20.3	22.7	10.6	–6.1
Other developed countries	11.2	13.4	14.7	0.8
Developing world	7.3	9.0	14.0	3.8
Developing world excluding China	8.1	9.4	9.2	–0.1
Developed world	22.6	24.3	21.7	–1.0
World	11.6	12.8	15.7	1.7

SOURCE: Calculated by authors from FAO statistical databases, www.apps.fao.org/subscriber (accessed January 2002).
NOTES: Data are three-year averages centered on 1973, 1985, and 1997, respectively. Growth rates are exponential, compounded annually using three-year averages as endpoints.

lower per capita consumption in the former Eastern Bloc countries.

Fish are an important source of protein, especially in developing countries. Fish account for 20 percent of animal-derived protein in low-income, food-deficit countries, compared with 13 percent in the industrialized countries.[5] Still, despite the rapid growth of fish consumption in developing countries, the level of per capita consumption was much lower than in the developed countries in 1997 (Table 1). The expansion of fish demand in developing countries is likely to continue for some time.

Consumption of freshwater fish (such as carp) and fish that migrate between freshwater and saltwater (such as salmon) has shown the greatest increase over recent years, especially in China, where per capita consumption of these fish increased nearly 10-fold between 1981 and 1997. Global per capita consumption of crustaceans has nearly tripled since 1970, with shrimp, both farmed and wild-caught, leading the way. Per capita consumption of shellfish such as oysters and clams has also tripled. China had a 13-fold increase in per capita consumption of these mollusks from 1981 to 1997. Consumption of fish that are not suited to aquaculture production, such as many marine fish, has remained relatively constant because of the stagnation in wild fish production.

The changing profile of fish consumption around the world comes as no surprise, partly because countries with rapid population growth, rapid income growth, and urbanization tend to have the greatest increases in consumption of animal products, including fish products.[6] Whereas population has hardly grown in the developed world over the past 15 years, it has grown rapidly in the developing world (2.1 percent a year, excluding China). Moreover, urbanization tends to change people's diet preferences, driving increased fish consumption. Finally, as individuals become wealthier, they tend to substitute higher-priced calories for lower-priced ones, once they have met their basic food needs. Income growth and urbanization have both been major factors in China's increased fish consumption, for instance.

Demand for fish as feed is also on the rise. Nearly one-third of the world's wild-caught fish are "reduced" to fishmeal and fish oil, which are then used in feeds for terrestrial livestock and farmed carnivorous fish. Aquaculture's share of demand for fishmeal has grown significantly over the past decade and a half (Figure 2). Regions with rapidly growing poultry, pig, and aquaculture sectors, such as China and Southeast Asia, have increased their use of fishmeal and fish oil, both absolutely and as shares of global use. Because aquaculture is likely to grow quickly over the next 20 years, some experts are concerned that rising demand for fishmeal and fish oil could place heavier fishing pressure on already threatened stocks of fish used for feed.

Figure 2 Changing uses for fishmeal, 1988 and 2000

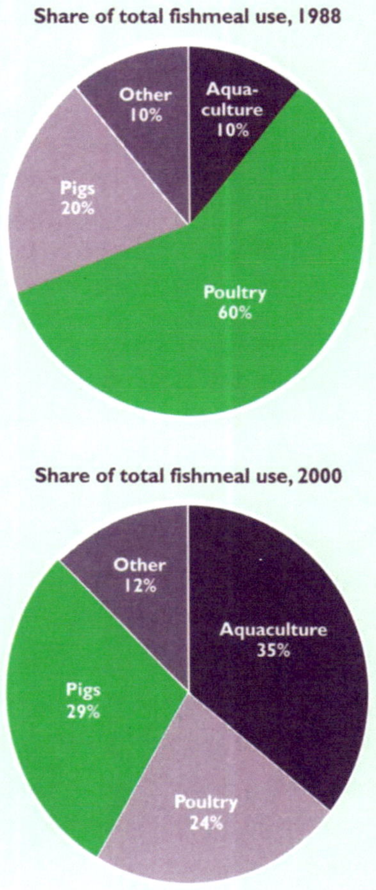

SOURCE: S. M. Barlow and I. H. Pike, "Sustainability of Fish Meal and Oil Supply," paper presented at the Scottish Norwegian Marine Fish Farming Conference, "Sustainable Future for Marine Fish Farming," University of Stirling, Stirling, Scotland, June 14–15, 2001.

The Shift of Wild Fish Production to Developing Countries

During the 1970s and 1980s exploitation of wild fish stocks soared, thanks to expanded fishing fleets, new fishing technologies, and increased investments in the fishing sector. Global capture of fish for food jumped from 44 million tons in 1973 to 65 million tons in 1997 (Table 2). By the late 1980s, however, the stocks fished by many wild-fishing operations were fully exploited and even overexploited. Since then, despite increases in investment and fishing capacity, fish production from wild fisheries has slowed or stagnated.

Since the 1980s, developing countries have taken the lead in producing fish from wild fisheries. Developing countries now account for more than 70 percent of the total production of fish for food,

Table 2 Production of wild fish for food, 1973–1997

REGION/COUNTRY	TOTAL PRODUCTION (MILLION METRIC TONS) 1973	1985	1997	ANNUAL GROWTH RATE 1985–97 (%)
China	3.8	5.0	13.9	8.9
Southeast Asia	5.0	6.9	10.4	3.5
India	1.7	2.1	2.9	2.8
Other South Asia	1.1	1.1	1.6	3.1
Latin America	2.3	4.1	5.7	2.9
West Asia and North Africa	0.7	1.4	2.1	3.3
Sub-Saharan Africa	2.1	2.6	3.7	3.1
United States	1.7	3.5	4.0	1.1
Japan	7.8	8.4	4.4	–5.2
European Union 15	5.6	4.9	4.7	–0.4
Eastern Europe and former Soviet Union	7.7	8.8	4.7	–5.1
Other developed countries	2.8	3.7	4.2	1.0
Developing world	18.9	26.9	42.5	3.9
Developing world excluding China	15.1	22.0	28.7	2.2
Developed world	25.6	29.3	22.0	–2.4
World	44.5	56.3	64.5	1.1

SOURCE: Calculated by authors from FAO, Fishstat Plus: Universal Software for Fishery Statistical Time Series (Rome: FAO Fisheries Department, Fishery Information, Data and Statistics Unit, 2002).
NOTES: Data are three-year averages centered on 1973, 1985, and 1997, respectively. Growth rates are exponential, compounded annually using three-year averages as endpoints.

including both wild fisheries and aquaculture (Figure 3 on the following page). Whereas developed-country production from wild fisheries exceeded developing-country production by 6.6 million tons in 1973, by 1997 the developing countries were producing twice as much as the developed countries. Part of this shift is due to the establishment of 200-mile exclusive economic zones (EEZs) within which coastal nations can claim exclusive fishing rights, excluding some developed-country fleets and forcing others to strike deals in order to gain fishing access. In some cases creation of these zones led to the reclassification of fish production from developed to developing countries, as vessels merely changed flags. Meanwhile, developing countries were expanding their own fishing fleets as developed countries were contracting theirs.

One of the most striking trends in the capture of fish for food has been China's emergence as the largest producer and the simultaneous decline of Japan's production. In 1973 Japan was the world's largest producer of wild food fish, accounting for 18 percent of global production. By 1997 its share had plummeted to 7 percent and its absolute level of production had dropped by nearly half. Enforcement of EEZs significantly reduced the fishery resources available to Japan, and dwindling stocks of fish such as pilchards further reduced Japanese catches.

Meanwhile, China increased its output from 9 percent to 21 percent, boosting production from under 4 million tons to 14 million tons. But China's astonishing growth during the 1990s in fish production, and the contrast between trends in China and in neighboring countries, has raised suspicions about the accuracy of reported totals. There is a significant and growing discrepancy between estimates of China's fish consumption based on independent household surveys and estimates of fish availability derived from production, trade, and other use data. Moreover, in the case of wild fisheries, reported catches have risen rapidly even though major stocks were classified as overexploited, and independent vessel survey data are at odds with Chinese estimates of catch. One study concluded that Chinese fishery production—including aquaculture—was overestimated by 43 percent in 1995 and suggested that institutional incentives that reward or punish local officials based on reported productivity may be largely responsible for the increasing distortion.[7] If China has indeed overreported its fish production for institutional or other reasons, trends in global fish production appear much less favorable to the health of stocks than they otherwise do.

Southeast Asian countries, especially Indonesia and Thailand, also dramatically increased their production of wild food fish, more than doubling output from 5.0 to 10.3 million tons between 1973 and 1997. More than one-quarter of the overall increase in wild fish production since 1985 can be attributed to the Indian Ocean, which in 1997 still represented less than 10 percent of global capture. The Indian Ocean was the sole major marine fishing area to show sustained growth in the production of wild fish in the past three decades.

Peru and Chile led Latin America's production, which also grew significantly, from 2.3 million tons in 1973 to 5.7 million tons in 1997. Production from Eastern Europe and the former Soviet Union declined precipitously after the fall of the Soviet Union, as the heavily subsidized Soviet and Eastern European fleets aged quickly and investment declined. Total production in these regions fell by nearly half from 1985 to 1997. European production declined over the same period, whereas production in Sub-Saharan Africa, West Asia and North Africa, the United States, and India

Figure 3 Changing share of developing countries in the production of fish for food, 1973 and 1997

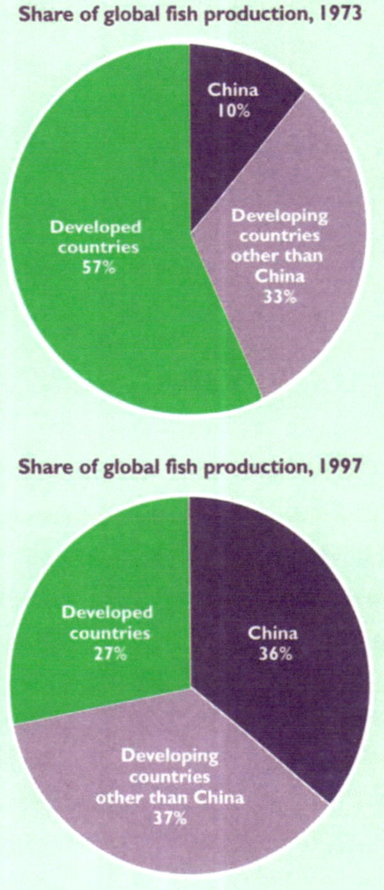

SOURCE: Calculated by authors from FAO, Fishstat Plus: Universal Software for Fishery Statistical Time Series (Rome: FAO Fisheries Department, Fishery Information, Data and Statistics Unit, 2002).

NOTE: Data are three-year averages centered on 1973 and 1997, respectively.

grew. Production from all Atlantic areas has stalled above 20 million tons since 1970. The high-profile collapse of the cod fishery industry in the northwest Atlantic has become emblematic of the threats posed by heavy fishing.

What are the prospects for future wild fish production? Forecasting production from wild stocks is an extraordinarily uncertain exercise. It is fairly clear that because most wild fisheries are near their maximum sustainable exploitation levels, production from these fisheries will likely grow only slowly to 2020. Although fishers could probably produce more by targeting under-exploited species that have been in lower demand (principally small mesopelagic species and krill), this strategy is untested in terms of consumer acceptance. More important, it could cause large shifts in species composition and indirectly harm predator species, with severe consequences for the environment. Large fishery stock collapses are also possible.

The Rising Share of Aquaculture

With wild fish production stagnating, growth in overall fish production has come almost entirely from the global boom in aquaculture, especially in developing countries. Aquaculture now represents more than 30 percent of total food fish production, up from just 7 percent in 1973. From 1985 to 1997 developing-country production of fish from aquaculture grew at an annual rate of 13.3 percent, whereas production in developed countries grew at a rate of 2.7 percent (Tables 3 and 4). Asia accounts for 87 percent of global aquaculture production by weight, and China alone commands a stunning 68 percent share, rising from 32 percent in 1973. Figure 4 shows China's increasing dominance in aquaculture production.

Aquaculture ranges from simple ponds using naturally occurring food sources to highly intensive systems with water control, aeration, and supplemental feeding. It is practiced inland, along the coast in brackish water systems, and in marine cages and net pens. Farm size can range from thousands of hectares down to the size of a backyard. The majority of global production comes from freshwater aquaculture (58 percent in 1999), followed by mariculture (36 percent) and brackish water (6 percent).

In the coming decades aquaculture will likely be the greatest source of increased fish production. Fish farmers can increase production through two channels: expanding the water surface area under cultivation or increasing yields per unit of area cultivated. To increase yields, they can either increase inputs or achieve greater efficiency from a given level of inputs. Both sources of yield growth are likely to contribute to aquaculture production in the next several decades.

Table 3 Annual growth rates of the production of fish for food, 1985–97 (%)

REGION/COUNTRY	ALL FOOD FISH	AQUACULTURE	CAPTURE
China	12.2	15.6	8.9
Developing world excluding China	3.0	8.4	2.2
Developing world	6.3	13.3	3.9
Developed world	-1.9	2.7	-2.4
World	3.1	11.2	1.1

SOURCE: Calculated by authors from FAO, Fishstat Plus: Universal Software for Fishery Statistical Time Series (Rome: FAO Fisheries Department, Fishery Information, Data and Statistics Unit, 2002).

NOTE: Growth rates are exponential, compounded annually using three-year averages as endpoints.

Table 4 Production of fish from aquaculture, 1973–97

REGION/COUNTRY	TOTAL PRODUCTION (MILLION METRIC TONS) 1973	1985	1997	ANNUAL GROWTH RATE 1985–97 (%)
China	1.0	3.4	19.5	15.6
Southeast Asia	0.4	0.9	2.3	7.6
India	0.2	0.6	1.9	9.6
Other South Asia	0.1	0.1	0.5	10.5
Latin America	...	0.1	0.7	19.4
West Asia and North Africa	...	0.1	0.2	9.2
Sub-Saharan Africa	...	...	...	11.7
United States	0.2	0.3	0.4	1.9
Japan	0.4	0.7	0.8	1.6
European Union 15	0.5	0.8	1.2	3.3
Eastern Europe and former Soviet Union	0.2	0.4	0.2	-6.4
Other developed countries	...	0.1	0.6	17.8
Developing world	1.8	5.7	25.4	13.3
Developing world excluding China	0.8	2.3	5.9	8.4
Developed world	1.3	2.3	3.2	2.7
World	3.1	8.0	28.6	11.2

SOURCE: Calculated by authors from FAO, Fishstat Plus: Universal Software for Fishery Statistical Time Series (Rome: FAO Fisheries Department, Fishery Information, Data and Statistics Unit, 2002).

NOTES: Data are three-year averages centered on 1973, 1985, and 1997, respectively. Growth rates are exponential, compounded annually using three-year averages as endpoints. An ellipsis (...) indicates quantities less than 0.1 mmt when rounded.

Figure 4 Total aquaculture production, 1970–99

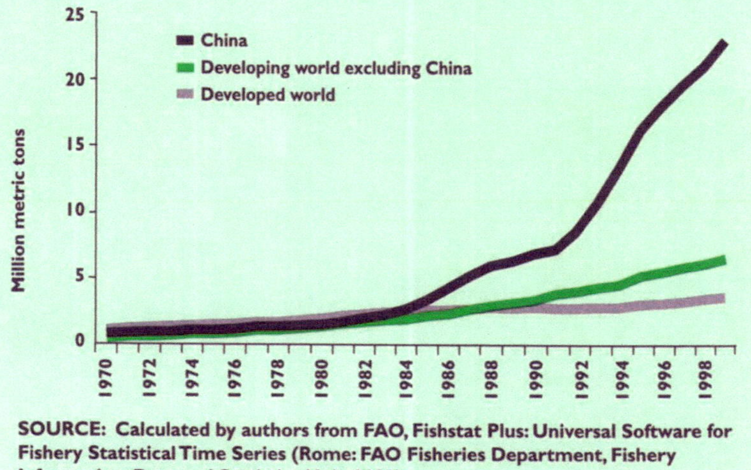

SOURCE: Calculated by authors from FAO, Fishstat Plus: Universal Software for Fishery Statistical Time Series (Rome: FAO Fisheries Department, Fishery Information, Data and Statistics Unit, 2002).

Because of the slow growth in wild fisheries, the level of aquaculture production will play a large role in determining the relative prices of fish commodities. Aquaculture's course is far from certain, however. The sector must overcome several major challenges if it is to sustain the rapid growth of the past 20 years. It will face competition for land and marine resource use from other activities ranging from terrestrial agriculture to recreation. Freshwater will become increasingly scarce over the next 20 years,[8] making further expansion of freshwater aquaculture more difficult.

Disease and lack of fishmeal and fish oil derived from wild-caught fish may also constrain aquaculture production. Selective breeding, better management of fish health, water control, and modification of feed inputs can help increase productivity. Growth in aquaculture production will depend heavily on the level of public and private investment in the sector.

The Turnaround in Fish Trade

Fish products are a heavily traded commodity, and the direction of trade is changing. Roughly 40 percent of global fish output by value in 1998 was traded across international borders (about 33 percent by weight),[9] compared with less than 10 percent of global meat output.[10] The high share of trade in fish is astounding for such a highly perishable commodity group. It reflects major changes in human diets around the world, changing supply infrastructure in both the North and the South, and the ongoing globalization of high-value food chains.

The enormous rise in fish production in developing countries has caused an about-face in the direction of trade in fish products since the early 1970s (Table 5). In 1973 the developed world was a net exporter of 818,000 tons of food fish, but by 1997 these countries were net importers of 4,045,000 tons of food fish. By the late 1990s more than 50 percent of fish exports came from developing countries, and two-fifths of developing-country fish exports originated in low-income, food-deficit countries. In 1999 net exports of fish and fish products from developing countries to developed countries surpassed US$16.5 billion.[11]

Food fish are primarily traded as fillets, packaged and cleaned frozen fish, and canned fish.[12] If fishmeal is included, more than 90 percent of fish trade is composed of commodities that have been processed in some form.

In 1997 high-value fish made up 90 percent of the gross fish exports of developing countries that were identified by species. Tuna trade, long a mainstay of international fish trade, is dominated by exports from China and Latin America and imports by the European Union, Japan, and the United States. Trade in cod continues to be overwhelmingly from North to South, but salmon exports have shifted from the United States to Latin America. Shrimp, which accounted for one-fifth of world fish trade by value in the late 1990s,[13] comes largely from Southeast Asia, Central America, and India and flows to the industrialized countries. Moreover, freshwater fish such as Nile perch, tilapia, and catfish have become significant export items from developing countries over the past 15 years. Sub-Saharan Africa's net exports of these fish rose by 53,000 tons from 1985 to 1997.

The Rising Price of Fish

The prices of many animal-origin foods have declined steeply over the past several decades because of increased production and stagnating demand in the traditional markets of the North. Real red meat prices, for instance, have declined by a stunning 50 percent since 1980. In sharp

Table 5 Net exports of fish for food, 1973 and 1997

	THOUSAND METRIC TONS	
REGION/COUNTRY	1973	1997
China	-108	181
Southeast Asia	-324	1,131
Latin America	44	2,435
Japan	520	-3,112
European Union 15	-989	-3,251
Developing world	-818	4,045
Developed world	818	-4,045

SOURCE: Calculated by authors from FAO statistical databases, www.apps.fao.org/subscriber (accessed January 2002).

NOTES: Negative values indicate net imports. Data are three-year averages centered on 1973, 1985, and 1997, respectively.

contrast, consumers have experienced a long-term increase in the real prices of fresh and frozen fish since World War II (Figure 5). Exceptions are canned finfish, which have become less favored in the consumption baskets of developed countries since the early 1970s, and some individual commodities like shrimp and salmon, which have seen large gains in production owing to aquaculture. The largest price increase during the period 1985–97 occurred for high-value finfish, probably as a result of rising demand and stagnating production.

Figure 5 U.S. producer price indexes for fish products, 1947–2000

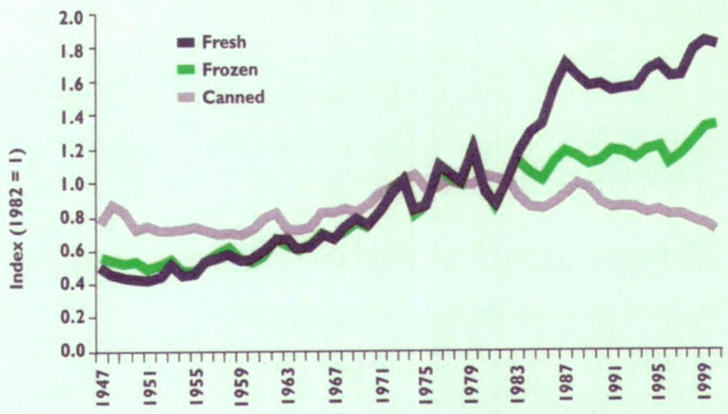

SOURCE: Calculated by authors from U.S. Bureau of Labor Statistics, Producer Price Index Commodity Data, www.data.bls.gov (accessed January 2002).

NOTE: Data were deflated by U.S. producer price index for all commodities.

Projections and Scenarios to 2020

To help analysts and policymakers understand the policy options that will confront them with regard to fisheries in the coming decades, this study projects supply, demand, and trade for fish to 2020. We draw on a tool called the International Model for Policy Analysis of Agricultural Commodities and Trade (IMPACT), developed and maintained by a team of IFPRI researchers led by Mark W. Rosegrant. IMPACT has been modified to include four categories of food fish: high-value finfish (such as salmon and tuna), low-value food fish (such as herring and carp), crustaceans (such as shrimp and crabs), and mollusks (such as clams and squid). It also includes two animal feed items made from fish: fishmeal and fish oil. It differentiates between fish produced from wild fisheries and those produced from aquaculture.

The model spells out the results for fish supply, demand, and trade under six scenarios: (1) a baseline scenario, using the most plausible set of assumptions about population and income growth, policy decisions, technology, and other factors; (2) faster aquaculture expansion, in which the rate of technological progress in aquaculture is 50 percent greater than under the baseline scenario; (3) lower Chinese production, in which figures for Chinese base production levels and growth are adjusted downward; (4) fishmeal and oil efficiency, in which the feed conversion efficiency of these two items increases twice as fast as in the baseline scenario; (5) slower aquaculture expansion, in which technological progress in aquaculture is half as fast as in the baseline scenario; and (6) ecological collapse, in which an exogenous declining trend of 1 percent annually is applied to production of all wild-fish commodities, including fishmeal and fish oil.[14]

Outlook for Fish Prices

The ultimate outcome of supply, demand, and trade patterns is manifested in the price of a commodity. Prices are the best indicator of incentives to both producers and consumers and have important implications for food security. Higher prices indicate relative scarcity and lessen the ability of consumers to purchase the commodity, while lower prices represent increased availability to consumers.

Fish are likely to continue to become more costly to consumers compared with other food products over the next two decades. In the baseline scenario, which is judged the most likely, real prices for fish commodities are projected to rise between 1997 and 2020 (Table 6). For high-value finfish and crustaceans, this price increase will be on the order of 15 percent. For fishmeal and fish oil, prices will rise even more—18 percent. Mollusks and low-value food fish are forecast to have significantly lower but still positive real price appreciation. These results are striking when compared with those for other food commodities, which show nearly uniform price declines. Fish will become about one-fifth more expensive relative to other kinds of meat by 2020. Fishmeal and fish oil will become 12 percent more expensive relative to low-value food fish, 19 percent more expensive relative to vegetable meals (substitutes for fishmeal and oil in animal feed), and 20 percent more expensive relative to poultry (for which fishmeal is an input).

In fact, fishmeal and fish oil prices would shoot up under several of the scenarios. The worst case would be an ecological collapse of wild fisheries, where the decline in fishmeal output coupled with increased demand for fishmeal from aquaculture would more than double prices by 2020. The faster aquaculture expansion scenario would also put significant upward pressure on fishmeal prices.

The one scenario that leads to slightly lower real fishmeal prices is the one that improves efficiency in fishmeal and fish oil conversion through

Table 6 Projected total change in prices under different scenarios, 1997–2020

	PROJECTED TOTAL CHANGE IN PRICES (%)					
COMMODITY	Most likely (baseline)	Faster aquaculture expansion	Lower China production	Fishmeal and fish oil efficiency	Slower aquaculture expansion	Ecological collapse
Low-value food fish	6	-12	6	5	25	35
High-value finfish	15	9	16	14	19	69
Crustaceans	16	4	19	15	26	70
Mollusks	4	-16	3	3	25	26
Fishmeal	18	42	21	-16	0	134
Fish oil	18	50	18	-5	-4	128
Beef	-3	-5	-3	-4	-2	1
Pigmeat	-3	-4	-2	-3	-1	4
Sheepmeat	-3	-5	-3	-3	-1	2
Poultry meat	-2	-5	-2	-3	0	7
Eggs	-3	-5	-3	-4	-2	3
Milk	-8	-10	-8	-9	-8	-5
Vegetable meals	-1	3	0	-7	-4	16

SOURCE: Projections for 2020 are from IFPRI's IMPACT model (July 2002).

rapid technological progress. This scenario suggests that the carnivorous aquaculture industry (which produces salmon and shrimp, for example, that require fish in their rations) may gain potentially high returns from investing in higher fishmeal and fish oil efficiency.

The faster aquaculture expansion scenario leads to a drop in the projected real prices of low-value food fish, though it also causes a significant rise in the price of fishmeal. This result reflects the different fisheries involved in supplying low-value food fish and fish for feed. It suggests that aquaculture supplies a large share of the low-value food fish consumed by the poor and that investing in improving the productivity and sustainability of low-value food fish aquaculture is a good way of making it more affordable to them. The slower aquaculture growth scenario results in significant real price increases for all food fish commodities.

Outlook for Fish Consumption

In the baseline scenario people in the developing world will increase their total consumption of both high- and low-value food fish, whereas total consumption will remain static in the developed world (Table 7). The rates hardly change if China is removed from the calculation, suggesting that this consumption increase is a widespread structural phenomenon driven by population growth, urbanization, and income growth. Per capita consumption is projected to grow in most of the developing world in the baseline, although it will remain unchanged in Sub-Saharan Africa and the developed world.

Under the ecological collapse scenario, per capita consumption only declines a small amount—from 17.1 kilograms under the baseline scenario to 14.2 kilograms. The reason is that sharp price increases under this scenario slow the decline of growth in wild fish production and lead to increased aquaculture output, in addition to reducing demand pressure.

The lower Chinese production scenario leads to a decline in per capita consumption of food fish of 1 kilogram, but mostly because of its effects on Chinese consumption. Altering the model parameters in response to the controversy over Chinese fish production levels makes a difference for Chinese consumption and production trends, but it has surprisingly little effect on consumption or production outside China or on world prices for fish.

Faster aquaculture expansion would lead to a 1.9-kilogram increase in global per capita consumption of food fish over the baseline scenario. The effect is twice as strong in developing as in developed countries, but it is significant in both.

Outlook for Fish Production

In the most likely (baseline) scenario, global food fish production is projected to grow by 40 percent from 1997, at an average annual rate of 1.5 percent (Table 7). The production share of the

Table 7 Projected growth rates for fish as food, 1997–2020

	ANNUAL GROWTH RATE (%)			
REGION/COUNTRY	**Total food fish consumption**	**Total food fish production**	**Wild production**	**Aquaculture production**
China	2.0	2.0	1.1	2.6
Developing world excluding China	1.9	1.6	1.0	3.6
Developing world	2.0	1.8	1.0	2.8
Developed world	0.2	0.4	0.1	2.1
World	1.5	1.5	0.7	2.8

SOURCES: Consumption data for 1997 are calculated by authors from FAO statistical databases, www.apps.fao.org/subscriber (accessed January 2002). Production data for 1997 are calculated by authors from FAO, Fishstat Plus: Universal Software for Fishery Statistical Time Series (Rome: FAO Fisheries Department, Fishery Information, Data and Statistics Unit, 2002). Projections for 2020 are from IFPRI's IMPACT model (July 2002).

developing countries rises from 73 percent in 1997 to 79 percent in 2020, and China accounts for about 5 of the 6 percentage points in the increase. This projection is consistent with trends observed in the past 30 years, including China's expansion into distant-water fisheries previously occupied by fleets from developed countries.

The share of aquaculture in worldwide fish production is projected to increase from 31 to 41 percent in 2020 in the baseline scenario. China's share of food fish production from aquaculture increases from 59 to 66 percent, and other developing countries' share of production from aquaculture increases from 17 to 27 percent. Finally, the share of low-value food fish in total food fish production is remarkably stable, at about 48 percent, although production of low-value food fish will come increasingly from aquaculture.

In the other scenarios, assumptions about technological change or increased investment in aquaculture are crucial for production results. There is a difference of 25 million tons between the faster aquaculture expansion and slower aquaculture expansion scenarios, neither of which incorporates outlandish assumptions about the trend rate of growth. Whether or not investment flows into aquaculture will have a tremendous impact on fish production outcomes.

Technology matters greatly. Improving the feed conversion efficiency of fishmeal reduces fishmeal production by 1 million tons compared with the baseline, a result that would reduce fishing pressure on fish used as feed. In the extremely pessimistic ecological collapse scenario, total food fish production surprisingly declines by only 17 percent, largely because producers respond to resulting major price increases for fish products by pursuing greater aquaculture production.

Outlook for Fish Trade

Developing countries became major net exporters of fish products in the late 1990s. Net exports from the developing world are projected to continue through 2020, though at a lower level than presently (Table 8). This is mainly because of rising domestic demand within developing countries for fish—as well as for other animal products—because of population growth, income growth, and urbanization. Although China, India, and Latin America are all projected to be net exporters in 2020 under the baseline scenario, only in Latin America are net exports forecast to represent a significant share of domestic production through 2020. In other developing regions, demand will continue to outstrip supply.

Developing countries, taken as a whole, will continue to be net importers of low-value food fish and net exporters of high-value food fish. Nonetheless, many developing regions will begin to import high-value items on a large scale. China, for instance, is projected to become a significant net importer of crustaceans by 2020. Although the model framework does not specifically predict the direction of trade flows between country groups, it seems likely from these results that South–South trade will increase in overall importance during the next two decades.

Table 8 Projected change in net exports of fish as food, 1997–2020

	THOUSAND METRIC TONS	
REGION/COUNTRY	**2020**	**Net change, 1997–2020**
China	543	362
Southeast Asia	482	-649
Latin America	3,047	612
Japan	-2,663	449
European Union 15	-2,443	808
Developing world	2,813	-1,232
Developed world	-2,813	1,232

SOURCES: Data for 1997 are calculated by authors from FAO statistical databases, www.apps.fao.org/subscriber (accessed January 2002). Projections for 2020 are from IFPRI's IMPACT model (July 2002).

NOTE: Negative numbers indicate net imports.

Fisheries and the Natural Environment

The health of the natural environment is essential to maintaining fish harvest levels in the face of increasing demand. Unfortunately, fishing activities around the world often cause significant, large-scale damage to the aquatic environment, imposing costs both on the fish sector itself and on other users. In terms of area or number of organisms affected, wild fisheries dwarf aquaculture as a source of negative environmental impacts. There is clearly a limit to the capacity of the world's oceans to supply wild stocks of fish, and wild fisheries may be approaching this limit, at least under current management regimes. Yet aquaculture, which many hoped could ease pressure on threatened wild stocks, has environmental problems of its own.

Wild Fisheries

Of all the environmental impacts caused by the fish sector, overfishing poses by far the greatest environmental threat. Wild fish stocks are notoriously difficult to manage for a variety of reasons, including complicated and uncertain access issues and the difficulty of assessing the state of this complex biological resource. Overinvestment in fishing and the resulting overcapacity have led to excessive exploitation of fish stocks, especially in developed countries. During the 1970s and 1980s fleet size increased twice as fast as fish harvests. Most stocks of wild fish today are classified as fully exploited, and an increasing number are overexploited, in decline, or in recovery.[15] Although supply of wild fish remains stagnant and wild fishing is becoming less profitable relative to other sectors, resources are only slowly moving out of the industry, with the result that persisting overcapitalization further endangers wild stocks.

Wild-fishing operations capture, kill, and discard a massive quantity of undersized fish, fish with other undesirable characteristics, and nontarget species. Captured nontarget species are known as bycatch. Although the majority of bycatch is marketable and is thus kept and sold, much of it is simply discarded. Global discarded bycatch of fish and other marine organisms is currently estimated at more than 20 million tons a year, nearly one-quarter of the world fish catch. These nontarget species face high risks to their reproductive capacity. Measures to increase the use of bycatch (as surimi or fishmeal, for example) might serve only to discourage the adoption of technologies designed to reduce bycatch levels.

Some fishing practices also destroy marine habitats. Bottom trawling—the dragging of weighted nets across the seafloor—leads to significant levels of bycatch and substantially disturbs seafloor ecosystems. Blast fishing and poison fishing have had devastating effects on coral reefs in the Indo-Pacific and other regions. Damage to coral reefs is both biologically and economically harmful. Coral reefs are extremely productive and diverse habitats, and many coastal communities depend heavily on them for providing fish, tourism, coastal protection, and other benefits.[16]

Fisheries may also have sizable indirect effects on ecosystems. Removing massive quantities of a species necessarily leads to wholesale changes in the food web dynamics of that system. Fisheries have caused documented ripple effects on many levels of the food chain, reducing the abundance of seabirds and marine mammals on the higher end and of sea urchins and algae on the lower end.[17] Over the past few decades wild fisheries have increasingly turned to small pelagic fish that are lower on the food chain.[18] This shift suggests that the extraction patterns of wild fisheries may not be sustainable.

Aquaculture

The expansion and intensification of aquaculture production has been accompanied by increased movements of live aquatic animals and products, making the accidental spread of disease more likely. High stocking densities, poor water quality, and poor seed quality can lead to outbreaks of disease, which then spread to other ponds through water exchange.[19] Disease can then lead to pond abandonment and land degradation.

Effluent from aquaculture ponds and pens is often released directly into surrounding waterways, causing pollution problems stemming from fertilizer, undigested feed, and biological waste in the water. This effluent can contribute to eutrophication of downstream waters, harm benthic communities, and cause other damage to water and soil quality.

The concentration and scale of aquaculture are also important for environmental impacts. With the development of large aggregations of densely situated farms, the chances increase that pollution might drive wholesale ecosystem changes. Rapidly increasing demand for fishmeal and fish oil for use as feed in aquaculture has also led to concern that the farming of carnivorous and omnivorous fish will place pressure on the wild pelagic stocks from which fishmeal and fish oil are derived.

Over the past few decades coastal aquaculture development, especially shrimp farming, has caused the destruction of hundreds of thousands of hectares of mangrove forests, which are crucial for filtering nutrients, cleansing water, and protecting ecosystems from floods and storms. Although mangrove conversion has slowed in recent years, the profitability of aquaculture still places pressure on remaining coastal forests.

In addition, farmed fish that escape into the wild can threaten native species by acting as predators, competing for food and habitat, or interbreeding and changing the genetic pools of wild organisms. Traits bred into farmed fish are often different from those that confer reproductive fitness in the wild, and interbreeding between escaped farmed fish and wild fish may result in the loss of important local adaptations. The risk is greatest for small populations that are already threatened. Concern over escaped species is likely to intensify in coming years as genetically modified fish are developed for aquaculture. Although no transgenic fish have yet been approved for commercialization, both developed and developing countries have tested transgenic farmed species ranging from shellfish to freshwater fish to marine fish.[20] Simulations have demonstrated the theoretical possibility of transgenic fish introducing a "Trojan horse" gene that entirely wipes out a native population.[21] Escaped fish are intrinsically harder to monitor and control than vegetable crops or terrestrial animals.

Fish Prices and the Environmental Effects of Aquaculture on Wild Fisheries

The overall effects of aquaculture on wild fisheries depend a great deal on what happens to fish prices. If fish from aquaculture and wild fisheries are substitutes for one another, then increasing production from aquaculture could lower fish prices. Although it is possible that lower prices would initially lead to greater fishing effort in the wild-fish sector owing to high fixed costs in fishing fleets, in the long run lower food fish prices are likely to reduce fishing effort and benefit the health of wild stocks.

IMPACT permits comparison of wild fisheries production under scenarios with differing rates of growth in aquaculture. Even though the scenarios for faster and slower aquaculture expansion result in a significant price differential for fish, the difference in wild production between the two scenarios is fairly small (about 4 percent of wild production) because of the generally low price-responsiveness specified for wild fisheries production. Rapid aquaculture growth will probably increase pressure on fish caught for feed, so an important environmental question will be the extent to which carnivorous aquaculture can find replacements for fishmeal and fish oil.

Fish and Poor People

Given the growing importance of developing countries in both production and consumption of global fish products, the fish sector would seem to offer important opportunities for people in poorer countries to increase their incomes by producing fish and to improve and diversify their diets by further increasing their fish consumption. But poor people are facing new barriers in both areas.

Rising Barriers for Poor Fish Producers

Even by the standards of developing countries, landless fish workers and artisanal fishers are often among the poorest of people.[22] Employment in wild-fishing operations on boats belonging to others is traditionally a refuge from unemployment in other sectors in both developing and developed countries. Poor fishers in developing countries generally operate at a small scale and use traditional fishing practices. Unless policymakers take steps to protect these groups, several trends in the fish sector may threaten their livelihoods.

New technologies are likely to accelerate the intensification of aquaculture that has occurred in recent decades. Even environmental requirements are likely to contribute to capital-intensive production, because controlling negative environmental effects from aquaculture often requires expensive capital investments. If developing countries adopt subsidies similar to those in China and the industrialized countries, large-scale, capital-intensive operations are likely to emerge in developing countries at the expense of traditional and small-scale commercial fishers. These subsidies include cheaper land and credit and lower taxes and tariffs on imported inputs for large-scale operations.[23] Furthermore, in many developing countries the legal and institutional framework to promote or protect access rights for traditional fishers are either weak or not implemented.

Overall, aquaculture in Asia offers opportunities for modestly profitable fish farming for small landowners at low levels of intensification, uncertainty for landless laborers, and highly profitable commercial activity for larger-scale, capital-intensive operators with access to natural resources.

The rising importance of fish trade raises its own barriers to poor producers. Most fish trade is in processed products of some sort, and developed countries generally maintain higher tariff rates on processed fish commodities than on chilled fresh fish. Yet tariff rates for processed products are fairly low and are not a major constraint on the growth of fish exports from developing countries. Instead, the difficulty arises in nontariff barriers erected by developed countries in response to consumers' concerns about food safety related to, for instance, antibiotic residues, bacteria, heavy metals, and other contaminants in fish.

Developed countries have established new requirements for fish imports, including labeling requirements and Hazard Analysis and Critical Control Points (HACCP) plans, to alleviate consumers' concerns. Meeting these new requirements for documenting the safe handling, processing, and origin of fish products requires considerable experience, skill, and investment. The minimum installation cost in Bangladesh for an HACCP-certified plant ranges from US$270,000 to US$380,000; annual maintenance of an average small plant costs US$35,000 to US$70,000.[24] Bangladesh would need to spend 9.4 percent of its annual export sales of fish to install an HACCP plant and 1.3 percent to maintain it.[25]

The costs of compliance with process-based food safety systems such as HACCP are not likely to exclude developing countries from trade altogether, and are in fact both necessary to gain access to high-value markets and cheaper than older inspection-based systems. Thailand has some of the most efficient HACCP processes in Asia and processes one kilogram of fish at a cost of only US$0.10 to US$0.14 a year.[26] Economies of scale in implementing HACCP plans are likely, however, to make it difficult for small-scale operations to compete unless they are vertically coordinated with larger processing operations that can certify the compliance of their suppliers with safe processes.

Developing countries that can address new hygiene and food safety requirements, fair labor practices, and environmental needs will have the opportunity to capture more of the lucrative export market by pursuing better quality management at lower cost. But if the poor are to benefit from this potentially profitable activity, policymakers will need to find ways of including smaller-scale producers in these arrangements.

Increasing Prices and Poor Consumers

Fish provide high-quality, easily absorbable protein and a wide variety of vitamins and minerals.[27] Even a small amount of fish is an important dietary supplement for poor people who cannot easily afford animal protein and who rely mainly on starch diets. But over the past 30 years fish has become more expensive relative to other food items. Whereas the price of meat is half what it was in the early 1970s, the real prices of fish have not fallen. Prices for most fish items are firm because fish demand, primarily in developing countries themselves, is outstripping supply. And it is possible that the increasing globalization of fisheries and the rise of high-end fish exports from poor countries place upward pressure on low-value food fish prices, as producers switch focus to high-value export commodities. The rising cost of low-value food fish to the poor at present and the potential for further rises in the future are real policy concerns.

The Crucial Role of Technology in the Fish Sector

As demand for fisheries products grows during the next several years, technology must play a crucial role in the ability of suppliers to keep pace, by improving the management of wild fisheries while minimizing waste and raising the value of products through processing and handling, by minimizing the environmental impacts of aquaculture through more sustainable intensification, and by increasing the efficiency with which fishmeal and fish oil are converted into fish products in farmed fish.

Reducing Pressure on Wild Fisheries

Nearly one-third of the world's wild-caught fish are not consumed directly by humans but rather are "reduced" to fishmeal and fish oil and consumed in feed by farm-raised animals, such as chickens, pigs, and other fish. The bulk of wild fish destined for reduction are small pelagic fish species not generally consumed by humans, such as anchoveta and menhaden. So far, as aquaculture has consumed more fishmeal, the terrestrial livestock sector has consumed less, because poultry and livestock producers have increasingly switched to vegetable-based meals. But many aquaculturists have little choice about feed composition, because full substitution of vegetable products for fishmeal and fish oil is not yet possible in the diets of carnivorous fish and crustaceans. Studies have shown that when vegetable protein replaces fishmeal in feed, several aquatic species exhibit lower growth rates and higher mortality.[28]

This situation has raised concerns that demand for fishmeal and fish oil from the burgeoning aquaculture sector will raise prices for these commodities and place increasingly heavy pressure on wild fisheries to produce fish for feed. Certainly, aquaculture is already consuming a larger share of the fishmeal supply than it used to, and the price of fishmeal may be "decoupled" from that of its chief feed substitute, soymeal, during times of fishmeal scarcity (Figure 6). If fishmeal prices rise high enough, it might even become profitable to start processing low-value food fish into fishmeal, thus reducing a key source of animal protein to the poor.

Figure 6 Ratio of fishmeal price to soymeal price, January 1981–January 1999

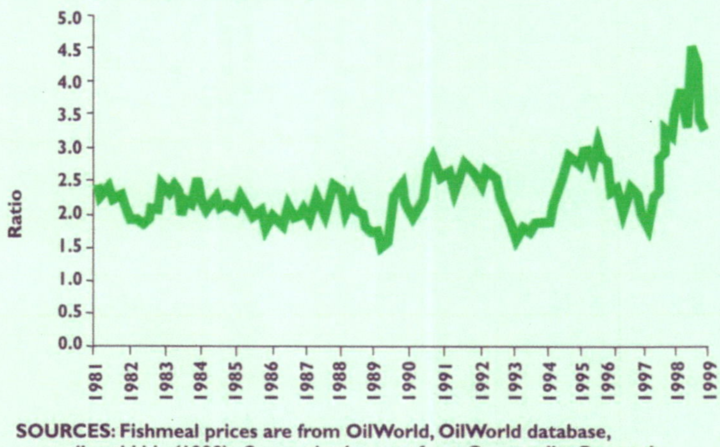

SOURCES: Fishmeal prices are from OilWorld, OilWorld database, www.oilworld.biz (1999). Soymeal prices are from Commodity Research Bureau, *CRB Commodity Yearbook* (New York: John Wiley and Associates, various years) through December 1997; and thereafter from U.S. Department of Agriculture Foreign Agricultural Service, *Oilseeds: World Markets and Trade, April 1999 circular*, www.fas.usda.gov/oilseeds/circular/1999/99-04/toc.htm (accessed June 1999).

NOTES: Fishmeal prices are CIF Hamburg. Soymeal prices are 44 percent protein at Decatur, Illinois, from January 1981 to September 1984; and 48 percent protein at Decatur, Illinois, from October 1984 to January 1999.

Technology can reduce the risks of higher prices and overfishing by providing alternatives to fishmeal and fish oil in aquafeeds. The share of fishmeal in aquafeeds has declined somewhat in recent years; some progress has been made in partially substituting protein-rich oilseed and grain byproduct meals for fishmeal in the diets of carnivorous finfish and marine shrimp. Other prospects for replacement include terrestrial animal byproduct meals such as meat-and-bone meal, although these may bring with them real and perceived risks for the spread of disease.

The main factor limiting the replacement of fishmeal with vegetable meals is the presence of elements within vegetable meals that inhibit nutrition for carnivorous fish species. Technology could help minimize their effects through either genetic selection of the cultivated fish species or through the use of improved feed-processing techniques.[29] Vegetable substitutes must also address concerns such as flavor, appearance, and nutritional content. Plant breeders have long modified soybeans and other crops to produce feed ingredients with more favorable commercial qualities, both by removing antinutritional factors and by increasing the content of desirable proteins. Genetic modification will increasingly be used to make further changes to existing vegetable crops. Totally replacing fish oil with commercially available plant and animal oil is even more problematic than replacing fishmeal. For many carnivorous fish species, fish oils are the only readily available source of essential fatty acids for these species, but research to address this issue is underway.

Improving Management of Wild Fisheries

The use of technologies such as nylon fishing nets, hydraulic power, and electronic fish-finding equipment led to the rapid growth of production from the 1950s to the 1970s. Now technological advances that improve information and management methods are needed more than advances to increase fishing capacity. Satellite remote sensing and other information technologies can help provide better information about the location, size, structure, and growth potential of stocks as well as help monitor fishing activity and improve consumer information about the condition and origin of fish products.

Even if information were perfect, the problem of effective management of fish stocks would remain. Access rights to fishing areas are subject to considerable regional and national variation and often lead to suboptimal outcomes in which fishers place excess pressure upon stocks. The creation of exclusive user rights, in which publicly owned and used resources are converted into publicly owned but privately used resources, may lead to increased cooperation among user groups and their acceptance of some of the responsibilities for management.[30] In other cases decentralized stakeholder management and collective action may be better suited to the governance of wild fishing. In any case, successful management of the world's wild-fishing operations will depend on the coordination of technology and policy. One example is a vessel monitoring system, which employs satellite tracking to allow onshore tracking of vessel movements, thereby enhancing the enforcement of regulations.

Technology is also crucial to avoiding the environmental damage and waste caused by certain fishing practices. Although some types of fishing gear may be banned altogether, others may be modified. Bycatch reduction devices, or BRDs, are increasingly used in fishing operations to lower the amount of unintended catch. BRDs may be designed to specifically exclude marine mammals, turtles, undersized fish, or other organisms. BRDs employed in Australian prawn trawls, for instance, have reduced bycatch by more than 60 percent while increasing the average size of prawns caught.[31] But without policy incentives to encourage their use, along with training and extension, BRDs will remain unused or ineffectively used.

Raising Productivity in Aquaculture

Compared with the advances achieved in the production of terrestrial animals, breeding technology in aquaculture is in its relative infancy. Significant productivity increases have been achieved for a few commercial species such as salmon, trout, and tilapia. The successful cultivation and breeding of species such as bluefin tuna, a lucrative and endangered fish that until now has only been wild-caught or grown without producing viable offspring, would be a tremendous boost to high-value aquaculture.[32]

Genetic modification and biotechnology also hold tremendous potential to improve the quality and quantity of fish reared in aquaculture, although not without significant controversy and risk. Biotechnology has the potential to enhance reproduction and the early developmental success of cultured organisms. Improved feed conversion efficiency in genetically altered fish would reduce the amount of feed inputs and waste per unit of output, possibly placing less pressure on the environment. Better growth and survival rates of cultured fish could reduce production costs per unit of output, possibly bringing down the price of fish to consumers.[33]

The possible environmental effects of genetically modified aquatic organisms are not well understood, and concerns exist over possible human health risks. The documented escapes of farmed salmon and their threat to native wild populations through interbreeding and competition demonstrate that caution should be employed when considering the introduction of a new species into an ecosystem. Although responsible development of improved fish in aquaculture requires a strong regulatory environment for monitoring and enforcement, this prerequisite is often lacking in both developing and developed countries. The successful adoption of transgenic technology in aquaculture will also depend upon consumers' acceptance of the new products.

Intensifying Aquaculture Sustainably

As demand for aquaculture products grows and investment levels rise, the value of cultivated area rises relative to labor and purchased inputs, provoking producers to replace land with labor and purchased inputs. As a consequence, aquaculture—particularly high-value aquaculture—has intensified over time. Because large-scale and intensive systems use higher levels of inputs and often generate high levels of outputs, they can potentially generate high levels of environmental problems. But intensification of aquaculture can also have positive effects on the environment.[34] Capital-intensive production systems often allow for more control over negative impacts such as effluent pollution and the spread of disease. Technology may in fact present economies of scale in the control of environmental problems.

Intensification can raise the risk of disease. Management techniques such as rotation of cultured species and lower-density stocking of organisms can partially address this risk, but antibiotics and water control technologies like aerators and water recirculation systems can also mitigate the stress caused by high concentrations of organisms. Advances in hatchery technology and the development of infrastructure to improve control over seed stock for shrimp would also significantly lower disease risks.

Effluent from aquaculture can raise problems both for the environment and for surrounding farms. Minimizing water exchange through recirculation has the dual benefit of reducing water demand and minimizing the effluent problem.[35] Other steps to improve water quality include calibration of the amount of feed used in order to minimize waste, integrated systems that raise complementary organisms to reduce unwanted outputs, and capital improvements such as aerators and pumps. Capital-intensive solutions to the effluent problem such as containment and treatment are probably not economically viable for the majority of the world's producers.

There is potential for aquaculture development in underexploited water bodies, such as rice paddies, irrigation canals, reservoirs, and seasonal or perennial ponds in developing countries.[36] Technology for such expansion need not be based on intensive commercial operations; rather, basic principles of aquaculture can be applied and adapted to local knowledge systems and different political and cultural contexts.

Some technologies long employed in traditional aquaculture systems can also be useful in addressing concerns raised by water management, effluent control, disease control, and land use in intensified aquaculture. For instance, polyculture, in which several species of organisms are grown together, can help reduce both inputs and negative outputs of a system. Integrated agro-aquaculture has succeeded in providing both cash and dietary protein to low-income farmers in Asia, Latin America, and Sub-Saharan Africa.[37]

The Role of Policy in the Fish Sector

Policymakers have an important role to play in ensuring that the fish sector contributes to poverty reduction and environmental sustainability in developing countries. In both developing and developed countries, policymakers should take a number of steps related to fish production. They should facilitate institutions that can help improve the governance of marine and coastal resources and freshwater fisheries. They should develop transparent and process-based food safety systems for consumers and focus on the sources of pollution in fisheries that most endanger human health and sustainability. They should adopt science-based policies concerning genetically modified aquatic organisms and develop ways of enforcing sustainable aquaculture codes of conduct. To ensure the survival of small-scale producers, they should promote market information, certification, and extension systems for these groups and develop a legal code for participatory institutions for small-scale fisheries, such as marketing and input supply cooperatives. They should redirect subsidies presently going to increase marine fishing operations to improving resource management and information systems. Finally, they should create a monitoring and planning function for fisheries policy within ministries of finance or prime ministerial offices to ensure that the sector gets the policy attention it deserves.

To improve policy outcomes in the developing countries, policymakers in the developed countries need to take a number of steps related to fish consumption and trade. They should rationalize food safety systems for seafood imports to promote safety and eliminate purely protective tariff and nontariff barriers. Tariff classifications should be harmonized and modernized across countries to fit the realities of fish trade. Policymakers should offer technical assistance to associations of small-scale, developing-country fish exporters for achieving fair trade and eco-labeling certification and for setting up and maintaining credible process-based food safety plans.

Finally, the focus of demand-side policies in developing countries should be to facilitate South–South trade, to provide public goods to assure domestic food safety, and to help ensure that fish products reach those in developing countries who need them the most from a nutritional standpoint. Thus policymakers should adhere to a rationalized rule-based system of fish trade that protects food safety but does not create artificial barriers to imports from other developing countries. They should use aquaculture as a development vehicle for the poor in regional development projects and then monitor and evaluate the impact of such projects. They should also monitor the nutritional status of poor people in fish-producing areas and explore cost-effective ways to improve outcomes using locally produced fish.

Conclusion

This assessment of the outlook to 2020 for fish in relation to developing countries, aquaculture development, poverty alleviation, and environmental sustainability points to five major structural shifts. These "sea changes" are already underway, though more visible in some cases than in others. By 2020, they will be pervasive. Forward-looking policy discourse, research, and technology development should focus on these changes.

First, the developing countries, particularly in Asia, will dominate production systems. To be sure, aquaculture development is central to this shift, but it will become more apparent in wild fisheries as well. The remaining one-quarter of world marine wild fisheries that are not fully exploited (all of which are in the tropics) will become more heavily fished.

Second, the source of net fish exports on a global scale has already shifted from the North to the South, and South–South trade will become increasingly important with the further emergence of urban middle classes. Developed countries will continue to be large net importers, and their domestic producers will gradually leave the sector. Over time, it is likely that public policy in the North will increasingly favor import-friendly regimes for fish. On the other hand, it is quite possible that trade wars—perhaps based on both real and spurious food safety claims—will become more prominent in the South. Fish will become an increasingly high-value food commodity in relative terms, and trade is likely to continue to shift from low-grade and frozen whole fish to fresh fillets and the like.

Third, environmental controversy will continue in the fish sector but will change focus. Overfishing in marine areas will remain a huge concern. Environmental regulations and institutions motivated by sustainability concerns will rapidly become more prominent, starting in developed countries and spreading to developing countries. The exploitation of fish for feed and of the stocks preyed on by traditional marine food fish will become a more important policy issue. It seems likely that the relationship between pollution and food safety in the fish sector will be given much more attention in both the North and the South. If problems become worse, and as the consumer base for fish becomes larger and more wealthy, more attention will be given to the sources of pollution, such as dioxins, PCBs, and heavy metal residues that accumulate in fish, affecting both wild fishing and aquaculture (through fish for feed). These pollution sources are outside the fish sector (they include runoffs from agricultural chemicals, industrial dumping of heavy metals, and chemical-laden rain), and the interests behind the activities causing the pollution have typically been stronger than the constituency worried about them. This will change.

Fourth, fisheries technology development will also shift to meet new challenges. Technology to profitably reduce the fishmeal and fish oil requirements for carnivorous aquaculture are key, and efforts will be expanded by private-sector interests. Some efforts will be focused on fish, others on synthetic feeds, and still others on modification of crops used in aquafeeds. The private sector will also continue to seek technologies to lessen the environmental problems caused by intensifying large-scale aquaculture. The public sector will show increased interest in finding technologies to mitigate environmental problems associated with the intensification of small-scale pond aquaculture under tropical conditions, where technological solutions to environmental problems have so far remained elusive. Environmental and food-safety regulations that require capital-intensive approaches to compliance will receive increased scrutiny. In wild fisheries, information technologies for improved management will become increasingly important in both the North and the South but will pay off for public purposes only where the right form of institutional development accompanies use of the technology.

Fifth, and finally, institutional development in the fish sector will be a necessary condition for reducing poverty through the development of the fish industry, as it is for improving environmental sustainability and food safety. The outlook for traditional fishers in developing countries in the absence of such institutional innovation is not bright. Both wild and aquaculture fisheries are scaling up and becoming more capital intensive, and increased focus on food safety and environmental effects under current technologies is likely to further this tendency. Food safety certification will become important to the survival of all fishers in the next two decades, and eco-labeling will become important to most. The world has not yet found a way to deliver such certifications cost-effectively and credibly to large numbers of small-scale fish producers, but the stakes are increasingly clear.

Developing a sustainable fish sector that benefits poor people will require researchers to address a number of other questions: What is happening to the industrial organization of fish production and processing in developing countries and why? How does the changing structure of international fish markets affect the opportunities and constraints for developing-country exporters, and specifically small-scale and poor producers? In particular, what has been the impact of eco-labeling, fair trade, organic labeling, and food safety regulations? How has the fish export boom of the late 1980s and 1990s affected incomes and nutrition of the poor in developing countries, and what can be done to improve outcomes? How can participatory institutions of collective action be designed to allow small-scale fishers and farmers to participate in growing fish markets subject to increasing economies of scale owing to rising environmental and health restrictions? How can participatory and market-oriented institutions be designed to improve the governance of resources critical to maintaining and expanding fish production? How can research on reducing the use of fishmeal in world fisheries be best promoted, and how can the costs of that research be recovered? At what point will world fishmeal prices become de-linked from soymeal prices, and what are the implications of this for the industry and for consumers? What are the constraints and opportunities for expanding South–South trade in fish, and what are the options for improving outcomes for poverty reduction and environmental sustainability?

These questions will be high-priority areas for further inquiry in an area of food policy whose global importance is becoming increasingly apparent.

Notes

1. All tons in this report are metric tons.
2. FAO, *The State of World Fisheries and Aquaculture* (Rome: FAO, 1995, 1998, 2000).
3. Pew Oceans Commission, *America's Living Oceans: Charting a Course for Sea Change*, http://www.pewtrusts.com/pdf/env_pew_oceans_final_report.pdf (accessed June 2003).
4. R. L. Naylor, R. J. Goldburg, J. H. Primavera, N. Kautsky, M. C. M. Beveridge, J. Clay, C. Folke, J. Lubchenco, H. Mooney, and M. Troell, "Effect of Aquaculture on World Fish Supplies," *Nature* 405 (2000): 1017–1024.
5. "Low-income food-deficit countries" and "industrialized countries" are FAO categories.
6. A. N. Rae, "The Effects of Expenditure Growth and Urbanization on Food Consumption in East Asia: A Note on Animal Products," *Agricultural Economics* 18 (1998): 291–299; C. L. Delgado and C. Courbois, "Trade-offs among Fish, Meat, and Milk Demands in Developing Countries from the 1970s to the 1990s," Proceedings of the Biennial Meetings of the International Institute of Fisheries Economics and Trade, edited by A. Eide and T. Vassdal (Tromso, Norway: Norwegian College of Fishery Science, 1998).
7. F. Lu, "Aspects of the Differences between Fishery Production and Consumption," *Management World* 1998 (5): 165–171.
8. M. Rosegrant, X. Cai, and S. Cline, *World Water and Food to 2025: Dealing with Scarcity* (Washington, D.C.: International Food Policy Research Institute, 2002).
9. FAO, *Fisheries Statistics: Commodities 1999* (Rome: FAO Fisheries Department, Fishery Information, Data and Statistics Unit, 2001).
10. C. L. Delgado, M. Rosegrant, H. Steinfeld, S. Ehui, and C. Courbois, *Livestock to 2020: The Next Food Revolution*, 2020 Vision Discussion Paper 28 (Washington, D.C.: International Food Policy Research Institute, 1999).
11. Ibid.
12. FAO, *Fisheries Statistics*.
13. Ibid.
14. More details about the parameters of the model and the six scenarios can be found in the book on which this report is based, *Fish to 2020: Supply and Demand in Changing Global Markets*.
15. FAO, *The State of World Fisheries and Aquaculture* (Rome, 2000).
16. H. S. J. Cesar, "Coral Reefs: Their Functions, Threats, and Economic Value," in *The Economics of Coral Reefs*, edited by H. S. J. Cesar (Kalmar, Sweden: CORDIO, 2000).
17. M. J. Williams, "Transition in the Contribution of Living Aquatic Resources to Sustainable Food Security," in *Perspectives in Asian Fisheries*, edited by Sena S. De Silva (Makati City, Philippines: Asian Fisheries Society, 1996).

18. S. M. Garcia and C. Newton, "Current Situation, Trends, and Prospects in World Capture Fisheries," in *Global Trends: Fisheries Management*, edited by E. L. Pikitch, D. D. Huppert, and M. P. Sissenwine (Bethesda, Md., U.S.A.: American Fisheries Society, 1997); D. Pauly, V. Christensen, J. Dalsgaard, R. Froese, and F. Torres Jr., "Fishing Down Marine Food Webs," Science 279 (1998): 860–863.

19. Funge-Smith, J. S., and M. R. P. Briggs, "Nutrient Budgets in Intensive Shrimp Ponds: Implications for Sustainability," *Aquaculture* 164, nos. 1–4 (1998): 117–133.

20. FAO, *The State of World Fisheries and Aquaculture*, 2000.

21. W. M. Muir and R. D. Howard, "Possible Ecological Risks of Transgenic Organism Release When Transgenes Affect Mating Success: Sexual Selection and the Trojan Gene Hypothesis," Proceedings of the National Academy of Sciences of the United States of America 24 (1999): 13853–13856.

22. G. Kent, "Fisheries, Food Security, and the Poor," *Food Policy* 22, no. 5 (1998): 393–404; M. Ahmed and M. H. Lorica, "Improving Developing-Country Food Security through Aquaculture Development: Lessons from Asia," *Food Policy* 27 (2002): 125–141.

23. M. Ahmed, "Fish for the Poor under a Rising Global Demand and Changing Fishery Regime," *NAGA Supplement* (July–December 1997): 4–7.

24. M. M. Dey, M. Ahmed, K. M. Jahan, and M. A. Rab, "Analysis of Fish Trade Policies in Developing Asian Countries: Liberalization vs. Barriers," paper presented at the biennial meetings of the International Institute of Fisheries Economics and Trade, Wellington, New Zealand, August 19–23, 2002.

25. J. C. Cato and C. A. L. dos Santos, "Costs to Upgrade the Bangladesh Frozen Shrimp Processing Sector to Adequate Technical and Sanitary Standards and to Maintain a HACCP Program," in *The Economics of HACCP: Costs and Benefits*, edited by Laurian J. Unnevehr (St. Paul, Minn., U.S.A.: Eagan Press, 2000).

26. Dey et al., "Analysis of Fish Trade Policies in Developing Asian Countries."

27. S. H. Thilsted, N. Roos, and N. Hassan, "The Role of Small Indigenous Fish Species in Food and Nutrition Security in Bangladesh," *NAGA Supplement* (July–December 1997): 13–15.

28. C. Lim, P. H. Klesius, and W. Dominy, "Soyabean Products," *International Aquafeeds* 3 (1998): 17–23.

29. G. Francis, H. P. S. Makkar, and K. Becker, "Antinutritional Factors Present in Plant-Derived Alternative Fish Feed Ingredients and Their Effects in Fish," *Aquaculture* 199, nos. 3–4 (2001): 197–228.

30. FAO, *Review of the State of World Aquaculture*, FAO Fisheries Circular No. 886 (Rome, 1997).

31. M. K. Broadhurst, R. B. Larsen, S. J. Kennelly, and P. E. McShane, "Use and Success of Composite Square-Mesh Codends in Reducing Bycatch and in Improving Size-Selectivity of Prawns in Gulf St. Vincent, South Australia," *Fishery Bulletin* 97 (1999): 434–448.

32. C. Iioka, K. Kani, and H. Nhhala, "Present Status and Prospects of Technical Development of Tuna Sea-Farming," paper presented at the Mediterranean Agronomic Institute of Zaragoza/FAO Seminar on Mediterranean Marine Aquaculture Finfish Species Diversification, held in Zaragoza, Spain, May 24–28, 1999.

33. Pew Initiative on Food and Biotechnology, *Future Fish: Issues in Science and Regulation of Transgenic Fish*, http://pewagbiotech.org/research/fish/ (accessed June 2003).

34. U. Barg and M. J. Phillips, "Environment and Sustainability," in *Review of the State of World Aquaculture 1997*, FAO Fisheries Circular 886 (Rome: FAO, 1997).

35. C. E. Boyd and A. Gross, "Water Use and Conservation for Inland Aquaculture Ponds," *Fisheries Management and Ecology* 7 (2000): 55–63.

36. C. H. Fernando and M. Halwart, "Possibilities for the Integration of Fish Farming into Irrigation Systems," *Fisheries Management and Ecology* 7, nos. 1–2 (2000): 45–54.

37. M. Prein and M. Ahmed, "Integration of Aquaculture into Smallholder Farming Systems for Improved Food Security and Household Nutrition," *Food and Nutrition Bulletin* 21 (2000): 466–471.

Glossary

Aquaculture	The cultivation of food fish or shellfish under controlled conditions, including marine net pens, freshwater ponds, brackish-water ponds, and cages
Aquafeeds	Formulated feed provided to organisms raised in aquaculture
Baseline data	The starting dataset for economic projections, in this case a consistent set of production, consumption, and trade figures for 32 food and feed commodities in 36 country groups averaged over annual FAO observations for 1996–98
Baseline scenario	The "most likely" scenario in economic modeling, incorporating the authors' best estimate of model parameters combined with the baseline data
Benthic	Living in or on the bottom of a body of water (such as clams)
Bivalves	Mollusks with two-valved shells, such as clams or mussels
Bycatch	The inadvertent catch of organisms that were not specifically targeted by a fishing operation (for example, nontarget fish species, marine mammals, seabirds) that are either discarded (see "discards") or landed for commercial sale
Carnivorous aquaculture	The cultivation of aquatic organisms that require animal matter as part of their feed, such as salmon or shrimp
Discards	Fish that are thrown away after being caught, usually because of undesirable characteristics (the wrong species, unmarketable, undersized, and so on); a subset of bycatch
Eutrophication	The process of enrichment of a body of water by organic nutrients, causing algal growth and thus reduced dissolved oxygen content, often resulting in the deaths of other organisms
Feed conversion efficiency	A measure of the ability to convert animal feed into animal meat, expressed as the ratio of live weight gain to feed ingested
Finfish	Aquatic vertebrates in the superclass Pisces; "true" fish as opposed to shellfish
Fish oil	Usually a byproduct of the fishmeal manufacturing process, used for pharmaceuticals, fish feeds, and for direct human consumption
Fishmeal	Cooked, pressed, dried, and milled fish, usually small pelagic fish, used for animal feeds
Food fish	Fish caught for consumption as food (rather than feed)

Krill	Very small planktonic marine crustaceans
Mariculture	Aquaculture practiced in a marine environment
Mesopelagic	Living in the region of the ocean between depths of about 200 and 1,000 meters
Pelagic	Living in the open ocean, as opposed to near shore or on the sea bottom
Polyculture	The cultivation of several species of organism within the same system
Surimi	Processed fish used for imitation seafood, often for artificial crab-meat
Wild fisheries	Fishing operations that catch wild fish, either in freshwater or salt-water

Made in the USA
Monee, IL
05 November 2026

67379782R00021